THE DEVIL'S DELUSION

A DISCUSSION GUIDE

This discussion guide is designed to facilitate the use of David Berlinski's book *The Devil's Delusion: Atheism and Its Scientific Pretensions* in small groups, adult Sunday School classes, adult education seminars, and book discussion clubs. The guide contains brief summaries of each chapter followed by discussion questions for that chapter.

Permission is hereby granted to distribute and reproduce this guide in whole or in part for non-commercial educational use, provided that: (1) the original source is credited, and any copies display the web address for Discovery Institute's Center for Science and Culture (www. discovery.org/csc) or FaithandEvolution.Org (www.faithandevolution. org); and (2) the copies are distributed free of charge.

John and Sonja West, October 2009

THE DEVIL'S DELUSION

A DISCUSSION GUIDE

BY

JOHN AND SONJA WEST

SEATTLE   DISCOVERY INSTITUTE PRESS   2009

Description
This is a group discussion guide for David Berlinski's book *The Devil's Delusion: Atheism and Its Scientific Pretensions.*

Copyright Notice

Library Cataloging Data

The Devil's Delusion, A Discussion Guide by John and Sonja West
47 pages, 6 x 9 inches, 229 x 152 mm.
BISAC Subject: REL067030 RELIGION/Christian Theology/Apologetics
BISAC Subject: REL004000 RELIGION/Atheism
BISAC Subject: SCI080000 SCIENCE/Essays

ISBN-13: 978-0-9790141-5-4   ISBN-10: 0-9790141-5-8 (paperback)

Publisher Information

Discovery Institute Press, Discovery Institute, 208 Columbia St., Seattle, WA 98104. Internet: http://www. discovery.org/

Published in the United States of America on acid-free paper.
First Edition, First Printing. October 2009.

CONTENTS

PREFACE

1. Atheist writer Sam Harris says that his strongest critics are Christians "who are intolerant of criticism." How does Berlinski respond to this claim?

2. Why did Berlinski write his book?

3. For whom did Berlinski write his book?

Personal Notes on the Preface

Chapter 1
No Gods Before Me

Chapter Summary: In the past, most scientists described religion as a separate realm from science that science could neither endorse nor disprove. The contemporary rise of militant atheism is an attempt by atheists to claim that science and religion are in conflict, and that science proves that religion is false. They demand that science be recognized as the one system of belief that explains everything and therefore demands unqualified allegiance.

1. How has the stance of scientists toward religion changed in recent years according to Berlinski?

2. Berlinski quotes evolutionary biologist Massimo Pigliucci stating that science is "a much more humble enterprise than any religion or other ideology."

   (a) Does Berlinski agree? Why or why not?

   (b) Who do you think is right? Why?

Personal Notes on Chapter 1

CHAPTER 2
NIGHTS OF DOUBT

Chapter Summary: Religious conviction plays an important role in human life. Atheists claim that religion produces evil and that secularism is making society better and better. The evidence doesn't support that. The twentieth century has an extensive record of brutal secular regimes. Even atheists don't try to claim that Lenin, Stalin, Hitler, Mao or Pol Pot were religious leaders. Without God everything is permitted. And no one is willing to live with that.

1. Berlinski notes that ancient Arabs made major advances in our understanding of astronomy.

   (a) What motivated these advances in astronomy?

   (b) How does this example help answer the claim that if scientists have religious motives for their work, then their scientific discoveries and ideas are unscientific or illegitimate?

2. Abu Hamid Muhammad Al-Ghazâli (1058-1111) is one of Islam's most influential philosophers.

   (a) According to Berlinksi, what concern did Al-Ghazâli express about science?

(b) Do you think this concern is fair or valid? Why or why not?

(c) In your view, do statements by "new atheist" authors Dawkins, Hitchens, and Avalos on sexuality confirm or undercut the concern expressed by Al-Ghazâli? Why?

3. According to Berlinski, "[f]or scientists persuaded that there is no God, there is no finer pleasure than resounting the history of religious brutality and persecution."

(a) How would you respond to the argument that religion is bad because it has produced a large amount of suffering and evil in human history?

(b) According to Berlinski, how does the experience of the twentieth century provide a counter-argument to those who blame religion for most of the world's evils?

4. Fyodor Dostoyevsky's 1881 novel, *The Brothers Karamazov* probes deeply into issues such as sin and free will.

(a) What warning was issued in *The Brothers Karamazov?*

(b) How does Berlinski think this warning applies today?

(c) Do you agree or disagree? Why?

5. Evolutionary psychologist Steven Pinker has claimed that "something in modernity and its cultural institutions has made us nobler" and that "[o]n the scale of decades, comprehensive data... paint a shockingly happy picture."

(a)What is his evidence?

(b) How does Berlinski respond to Pinker's claim?

(c) Who do you think has the better argument? Why?

6. Berlinski asks, "Does something in the very nature of a secular society make the monstrous possible?"

(a) How would you answer this question?

(b) What is Belrinski's response to this question, and how convincing do you find it?

7. According to Berlinski, "[o]ne might think that in the dark panorama of wickedness, the Holocaust would above all other events give the scientific atheist pause."

    (a) Why?

8. According to historian Richard Weikart, how did science—biology in particular—feed the ideology of the Nazis?

9. (a) According to new atheist Sam Harris, who is to blame for Hitler's genocide of the Jews? What do you think of his argument?

    (b) Harris criticizes Jews for not abandoning their belief in God after the Holocaust. What is Berlinski's response?

    (c) How does Berlinski respond to Sam Harris's dismissal of the moral concerns raised by biology?

    (d) What do you think about Harris's position?

10. (a) According to Berlinski, what is the view of human nature presented by new atheists Richard Dawkins and Sam Harris?

(b) How does the view of Dawkins and Harris compare to the Biblical view of human nature? (See Psalm 14:1-3, Ecclesiastes 7:20, Romans 3:9-20 and 3:23.)

(c) How does the view of Dawkins and Harris compare to the evidence provided by historical experience and common sense?

11. (a) According to Berlinski, what are the three possible answers to the question "What makes the laws of moral life true"?

(b) Which answer do you think is best? Why?

(c) Is it possible to combine two of the options?

(d) Which answer is preferred by the scientific atheists?

12. According to Berlinski, what are the problems with believing that moral judgments are based either on nothing or simply on what the majority of people think?

13. For those who know the writings of C. S. Lewis, how do some of Berlinski's arguments echo Lewis's arguments in chapter one of his book *Mere Christianity*?

Personal Notes on Chapter 2

CHAPTER 3
HORSES DO NOT FLY

Chapter Summary: Secular scientists denounce religion for teaching the existence of things that cannot be seen. Science does that, too. Scientists have faith that one elegant theory will explain all their data; that theory is unseen. Sub-atomic particles, like neutrinos and quarks, can't be seen either and the evidence that they exist is not simple.

Scientific studies matter, but scientists' atheistic claims are philosophical. They rest on beliefs that cannot be measured physically so that the atheist's attacks on religion are equally attacks on atheism. The evidence for most human beliefs, including the ordinary beliefs of atheists, is not the same kind of evidence as that undergirding mathematical physics.

1. What division is Berlinski talking about when he writes, "No division cuts deeper in the United States—or the world—or provokes a greater sense of mutual unease"?

2. (a) According to Berlinski, how is the role played by faith similar in both science and religion?

   (b) How is the role of doubt similar in both science and religion?

3. According to Berlinski, one popular argument for the non-existence of God is "If God exists, then his existence is a scientific claim, no different in kind from the claim that there is tungsten to be found in Bermuda. We cannot have one set of standards for tungsten and another for the Deity."

(a) What do you think of this argument?

(b) What is Berlinski's response to this argument?

4. (a) What is "naturalism"?

(b) Why does Berlinski think it cannot be appealed to as a convincing objection to God's existence?

5. (a) What is "methodological naturalism" and what does new atheist Hector Avalos think it provides a foundation for?

(b) How does the work of Isaac Newton contradict Avalos's assertion?

6. (a) What is materialism, and how is it based on faith according to Berlinksi?

(b) How does materialism, if true, make atheism more plausible?

(c) Does the view of the world supplied by modern science fit easily with the view supplied by philosophical materialism according to Berlinski? Why or why not?

(d) What does Berlinski mean by comparing modern materialism to a barroom drinker?

7. (a) According to Berlinski, what is so special about the scientific method?

(b) Why do you think Berlinski makes this point in a chapter devoted to discussing arguments against the existence of God?

8. (a) What does Berlinski think of the argument that empirical science is the only way of knowing the truth about something?

(b) How did David Hume make this type of argument, and what is Berlinski's response?

9. (a) According to Berlinski, how do scientists try to shut down outside criticism?

(b) What is Berlinski's critique of this kind of mindset among scientists?

Personal Notes on Chapter 3

CHAPTER 4
THE CAUSE

Chapter Summary: The cosmological argument, most powerfully stated by Thomas Aquinas in the 13th century, says that effects have causes and there cannot be an infinite series of causes. Therefore, there must have been a first cause. Aquinas identified this first cause with God: both God and the first cause are themselves uncaused.

A form of the cosmological argument appeared unexpectedly in 20th century physical cosmology, which now believes that the universe came into existence as a result of the Big Bang. The universe had a beginning. The Big Bang is well-established, being supported by both observational data and theoretical data in the form of the field equations of Friedmann-Lemaître cosmology. The Big Bang is a singular event—infinite temperature, no distance between particles, the beginning of matter and time—in which physical laws break down. In the beginning God created the heavens and the earth expresses the concept of the Big Bang well; this is disturbing to atheists.

1.   (a) What is the cosmological argument?

(b) In which cultures has it been made according to Berlinski?

(c) Is the cosmological argument by itself an argument for the existence of God according to Berlinski? Why or why not?

2. (a) Who was Thomas Aquinas, and what was his cosmological argument for the existence of God?

(b) What does Berlinski think of Aquinas's formulation of the cosmological argument? Do you agree or disagree? Why?

(c) What does Berlinski think of Richard Dawkins' objection to Aquinas's argument?

3. What does Berlinski think of philosophical objections that seek to diminish the need for a first cause?

4. According to Berlinski, how have the philosophical debates over the cosmological argument "been overtaken by events"?

5. (a) What is the Big Bang?

(b) What are the two lines of inference that made the Big Bang "irresistible" according to Berlinski?

(c) What evidence since the early 1960s further confirms the Big Bang?

6. Leading scientists such as Fred Hoyle initially rejected the Big Bang.

   (a) Why did they do this and was it just because of the science?

   (b) What does the initial rejection of the Big Bang indicate about the openness of the modern scientific community to discoveries that may support belief in God?

   (c) What can we learn from this experience for today?

7. How does the Big Bang provide new support for the cosmological argument?

Personal Notes on Chapter 4

Chapter 5
The Reason

Chapter Summary: Aquinas argued that if the universe is contingent (it might not have existed), then at some point it did not exist. Something must have brought the universe into existence. There must be something that is not contingent, but necessary. That something is God. Aquinas's argument does not prove that the universe is not eternal, but even if the universe existed eternally, it might not have existed. Why does it exist?

Physicists know that light is a wave and is also a particle. As a wave, a light particle can apparently be in two places at once, but when measured it is in one place. No one can explain exactly how this can be. Physicists have tried several theories to explain how everything has emerged from nothing while avoiding a theistic explanation. Perhaps the wave function of light also applies to universes. Maybe there are many universes and this is only one. These theories are not supported by any evidence.

1. Explain the second cosmological argument.

2. How has quantum mechanics changed the scientific understanding of reality?

3. (a) According to Berlinski, how have scientists such as Victor Stenger, Peter Atkins, and others tried to draw on quantum mechanics to disprove God?

   (b) Have they been successful according to Berlinski? Why or why not?

4. According to Berlinski, what are the two aims of concepts like the "Sea of Indeterminate Potentiality"?

5. (a) What is the many-worlds interpretation of quantum mechanics?

   (b) What does Berlinski think of it?

6. (a) What are the views of Stephen Hawking in this area?

   (b) What does Berlinski mean when he writes that "[i]f what Hawking has described is not quite a circle in thought, it does appear to suggest an oblate spheroid."

CHAPTER 6
A PUT-UP JOB

Chapter Summary: The physical properties of the universe are exactly what they need to be for human life. A slightly different cosmological constant, or ratio of neutrons to protons, or ratio of the electromagnetic force to the gravitational force, or speed of light would a disaster for us. Why is the universe so precisely fine-tuned? The obvious answer is that it was designed that way, but a great many physicists don't want to believe that.

The Standard Model of physics explains three of the four forces in the universe. It is a great triumph, but incomplete. The theory of General Relativity is still needed to explain gravity, and the two theories haven't been reconciled. Looking for an over-arching explanation, physicists advance a theory called string theory. String theory equations indicate thousands of different theoretical universes with multiple additional dimensions, but there is no physical evidence that any of those universes exist. If there were multiple universes, which many physicists are now insisting must be so, the fact that ours is so precisely fine-tuned is just chance. But why is our universe, the one we actually know about, so exact?

1. Why have some scientists complained that the universe looks like a "put-up job"?

2. What is the standard answer offered by theology for why the universe looks like a "put-up job"?

3. (a) What is the "Standard Model," and does it explain why the universe looks like a "put-up job"? Why or why not?

   (b) What is string theory and was it successful according to Berlinski? Why or why not?

4. According to Berlinski, how have scientists continued to try to avoid the obvious implications of the universe's fine-tuning through the Landscape and Anthropic principles?

5. (a) How does Berlinski use the account about Elijah in 1 Kings 19:1-8 to respond to the claim that the Anthropic Principle has explained away why our universe is fine-tuned for life?

   (b) Do you agree with Berlinski's point? Why or why not?

6. (a) According to Berlinski, what are the three possible answers for why an electron continues to follow the laws of nature?

(b) What is the answer preferred by the scientific atheists?

(c) Which answer do you prefer? Why?

Personal Notes on Chapter 6

Chapter 7
A Curious Proof that God Does Not Exist

Chapter Summary: Dawkins's argument against God acknowledges that the universe is improbable, but claims that God is also improbable and therefore doesn't exist. Since some atheists claim that the universe just happens to exist, even though improbable, it is equally valid for theists to claim that God just exists, even if He is improbable. If God created the universe, then the universe is not improbable.

Probability assigns numbers to events; things produced by chance are unlikely to be highly specified, but things that are designed show high specificity because they were designed that way. Dawkins's argument that something that is improbable doesn't exist is not a logically conclusive argument. Improbable only means "may not happen." It doesn't mean "cannot happen." Improbable events do happen. Dawkins's claim that God is improbable is ambiguous. Probability assigns numbers to a process, but Dawkins says nothing about whatever process he is anticipating would produce the deity. Aquinas said that something must exist that doesn't have a cause or nothing could exist. That something is necessary.

1. (a) What is Richard Dawkins' curious proof that God does not exist?

(b) What is David Berlinski's critique of Dawkins' argument?

2. (a) According to Berlinski, what is the underlying "power assumption" underlying Dawkins' argument and similar arguments?

(b) How does Berlinski use Sherlock Holmes as well as *The Perfect Storm* to rebut this underlying assumption?

3. In the end, how are materialist scientists like Steven Weinberg "inadvertent theists" according to Berlinski?

4. Berlinski writes at the end of the chapter: "What a man rejects as distasteful must always be measured against what he is prepared eagerly to swallow." How does he apply this standard to Richard Dawkins?

Personal Notes on Chapter 7

CHAPTER 8
OUR INNER APE, A DARLING, AND THE HUMAN MIND

Chapter Summary: With regard to human beings, the scientific community tries to assert that there is no significant difference between human beings and apes. Alfred Wallace, the co-creator with Charles Darwin of the modern theory of evolution, came to doubt the theory because of certain features of human beings. Both human physical features and features of the human mind cannot be explained in Darwinian terms. There are similarities between human beings and animals, but the differences are huge: language, art, architecture, music, dance, and mathematics are only a very short list of the areas of difference. While the similarity of the genetic code between human beings and chimpanzees is high, the function of the code is controlled by a series of regulatory systems both amazing complex and poorly understood.

Another attempt to explain away human exceptionalism is that human behavior today is the result of survival tactics of ancient humans persisting as a result of evolutionary biology. These explanations are implausible, unverifiable and irrelevant. Another attempt to reduce human beings is the claim that the human mind is only a machine. No machine does what the human mind does: react, respond, form intentions, take action. Evolutionary scientists do not take their own theories seriously with regard to themselves. They don't believe their thoughts or actions are determined by their genes, but if not then evolutionary psychology is irrelevant.

1. What is the basis of the idea that human beings have powers and properties that are unique in the animal kingdom?

2. (a) According to an editorial in the top science journal *Nature*, what idea "can surely be put aside?"

   (b) Do you agree? Why or why not?

   (c) What, if anything, does this editorial indicate about the attitude and views of those at the top of the scientific community?

3. (a) Who was Alfred Wallace?

   (b) Why has he been neglected by historians?

   (c) Why did Wallace think that evolution was inadequate in explaining the development of certain features of the human race? Which features?

   (d) What "forbidden doctrine" was suggested by the evidence Wallace presented?

For more information about the interesting views of Alfred Wallace, you can consult Alfred Russel Wallace's *Theory of Intelligent Evolution* (Erasmus Press, 2008), edited by Michael Flannery.

4. According to evolutionary biologist Frans de Waal "[i]f an extraterrestrial were to visit earth, he would have a hard time seeing most of the differences we treasure between ourselves and the apes."

    (a) Do you agree or disagree? Why?

    (b) How does Berlinski respond to de Waal's claim and the more general claim that human beings are essentially apes because they share a large part of their genome with chimpanzees?

5. (a) What is evolutionary psychology, and what claim does it advance?

    (b) Why does Berlinski think that evolutionary psychology has "no scientific value," indeed, "no value whatsoever"? Do you agree or disagree? Why?

6. (a) What is the argument made by evolutionary psychologist Steven Pinker and others about the human mind?

(b) What is Berlinski's critique of these arguments?

7.   At the end of the chapter, Berlinski writes: "One of the curiosities about the current enthusiasm for various peudo-scientific accounts of the human mind is that deep down those most willing to promote its premises are least willing to accept its conclusions." What does he mean?

Personal Notes on Chapter 8

CHAPTER 9
MIRACLES IN OUR TIME

Chapter Summary: We see truths about the physical universe with astounding accuracy, but our understanding is incomplete. Atheists claim that these gaps lead to the notion of a God of the Gaps, who will disappear when all the gaps are filled. The more scientists discover, though, the more gaps arise.

Darwinian biology is the only scientific theory both widely championed by the scientific community and widely disbelieved by the general public. The theory makes little sense and is supported by little evidence. Geographical barriers cause species to diverge, except that they also diverge where there are no geographical barriers. The fossil record shows sudden appearance of new species, not the gradual change that Darwin's theory demands. Species can be bred for traits, but the essential nature of the animal doesn't change. Biological research calls into question the very existence of natural selection.

Human and animal traits are often innate, giving rise to the claim that these traits are genetic. But genes are chemicals. The actions and intentions of organisms are not chemicals. If there is a connection between genes and human abilities, such as language, we don't know what that connection is. We have no explanation about the origin of life from non-life. We do not know how the process of light striking the eye and being converted to electrical signals in the brain produces our experience of vision. The things we embody as human beings, and our experiences of longing and love and anxiety and joy, are ignored by the sciences. In the deepest human sense, the things science explains don't matter.

1. Christopher Hitchens writes that "the age of miracles seems to lie somewhere in our past." Do you agree? Why or why not?

2. Critics of Darwin's theory are commonly accused of promoting a kind of argument from ignorance known as the "God of the Gaps." This argument claims that a lack of scientific knowledge for some natural phenomenon (a "gap" in our knowledge) means that "God did it." Supporters of Darwin's theory typically insist that "God of the Gaps" reasoning is fallacious because science has a strong record of "filling in the gaps" and providing material explanations for many phenomena that were previously mysterious.

   (a) Does Berlinski think this criticism is valid or invalid? Why?

   (b) It is sometimes claimed that the argument for intelligent design in nature represents an "argument from ignorance" or a "God of the Gaps" type of explanation. If you are interested in pursuing this objection, read Stephen Meyer's essay, "A Scientific History and Philosophical Defense of the Theory of Intelligent Design," available for free at http://www.discovery.org/a/7471. In the section titled "An Argument from Knowledge" (pp. 27-28), how does Meyer argue that intelligent design is not an argument from ignorance?

3. According to Berlinski, what are two reasons suspicions arise about Darwin's theory?

4. Explain and discuss Berlinski's view of the evidence for Darwin's theory provided by (a) the fossil record, (b) laboratory demonstrations of speciation, (c) natural selection, and (d) computer simulations.

5. Berlinski writes that "[i]f Darwin's theory of evolution has little to contribute to the content of the sciences, it has much to offer their ideology." What does he mean?

6. According to Berlinski, are evolutionary biologists in private as dogmatic as they are in public about the evidence for Darwin's theory? Why do you think this is?

7. (a) According to Berlinski, what challenges to Darwin's theory are presented in the paper "The Biological Big Bang Model for the Major Transitions in Evolution" by Eugene Koonin? (Koonin's paper can be accessed for free at http://www.biology-direct.com/content/2/1/21.)

    (b) What was the reaction of defenders of Darwin's theory to Koonin's paper according to Berlinski?

    (c) What does this reaction reveal about the mindset of many of Darwin's defenders?

8. How have the ideas of Japanese biologist Motoo Kimura challenged Darwin's theory of natural selection?

9. What is evolutionary biologist Michael Lynch's criticism of Richard Dawkins?

10. What problem has "eluded analysis" thus far according to evolutionary biologist Emile Zuckerkandl?

11. Given the views expressed by Koonin, Kimuri, Lynch, and Zuckerkandl, how should one regard Daniel Dennett's claim that natural selection has been demonstrated "beyond a reasonable doubt" or Steven Pinker's assurance that "[n]atural selection is the only explanation we have of how complex life can evolve"?

    If you want to pursue further discussion about the scientific evidence for and against Darwin's theory, check out the opposing articles available at http://www.faithandevolution.org/debates/was-darwin-wrong.php.

12. In the chapter section titled "Great Gaps of God," Berlinsk cites Job 38:4 ("Where was thou when I laid the foundations of the earth? declare, if thou hast understanding.")

    (a) Read Job 38:1-42:6. What is the main message of this passage?

(b) In the chapter sections "Great Gaps of God" and "Time, Death, Life, and Longing," how does Berlinski argue that modern science basically leads us to a similar message?

Personal Notes on Chapter 9

CHAPTER 10
THE CARDINAL AND
HIS CATHEDRAL

Chapter Summary: Galileo asserted that the Bible may be incorrectly interpreted. The natural world never transgresses God's law imposed on her. The grand book of the universe is always open to read, but people have to learn to read it.

The debate about the earth moving around the sun came to Cardinal Bellarmine. He said it was dangerous to say that the sun stood still and the earth moved around it. If we had evidence that the sun stands at the center, which we do not, the cardinal wrote, we would have to cautiously interpret the Scriptures which seem to deny that. But if the Scripture and nature were in fact found to be in conflict, it would be better to say we don't understand the Scripture than to say that what we know to be true is false.

Today, the scientific "scripture" is Darwinian theory and many scientists are willing to pursue a modern inquisition to silence those who question it. The faith that science, through its great physical theories, will be able to explain all of reality is the cathedral of our modern world. But science does not explain all of reality.

1. Galileo is one of the most important scientists in history. He was also a devout Catholic, and thus an interesting illustration of the relationship between science and religion.

    (a) According to Berlinski, what revolutionary doctrine did Galileo embrace with regard to the Bible?

(b) According to Berlinski, did Galileo really abandon the idea of inerrancy? Why or why not?

2. Berlinski suggests that the idea of the earth revolving around the sun poses a problem for those who believe in Biblical inerrancy. Theologian and astrophysicist Robert C. Newman disagreed:

> [W]hat about those passages in Scripture which seem to indicate that the earth is standing still? The strongest passages of this sort are those that speak of the sun rising, or the sun and moon standing still at Joshua's command... All of these passages are being misinterpreted when we read them to mean that the earth cannot be going around the sun. The Bible writers are speaking from a reference frame located at some point on the surface of the earth, and this is exactly the case with modern astronomers when they speak of the time of sunrise at Philadelphia. We no more need to fear that the One who inspired Scripture is making a mistake in the one case than are the astronomers in the other.

Who do you think is right on this point? Berlinski or Newman? Why?

3. Who was Niccolò Lorini, and what was his view of Galileo and his ideas?

4 Who was Cardinal Bellarmine, and what was his view of Galileo and his ideas?

5. According to Berlinski, how does the scientific worldview of today hold a similar position to that of the Bible at the time of Galileo?

6. According to Berlinski, who are today's Niccolò Lorinis?

7. How is the controversy over Stephen Meyer's paper in the Proceedings of the Biological Society of Washington an example of what Berlinski is talking about? (Additional information about this controversy can be found at http://www.richardsternberg.net/smithsonian.php.)

8 In the final section of the book, Berlinski offers a parable about the building of a great cathedral.

   (a) What does the cathedral in the parable represent?

   (b) What are the problems confronting the builders of the cathedral in the parable?

   (c) What truths do these problems in the parable represent for the scientific worldview?

(d) What is the viewpoint of the Cardinal Bellarmine character in the parable? Whose view is the Cardinal supposed to represent in this story?

(e) What do you think Berlinski's ultimate point is in this parable?

9. What do you think Berlinski's ultimate point is in his book?

10 What is the most important truth or insight you learned from reading and discussing this book?

Personal Notes on Chapter 10

www.ingramcontent.com/pod-product-compliance
Lightning Source LLC
Chambersburg PA
CBHW030307030426
42337CB00012B/624